YOUR KNOWLEDGE HAS VALUE

Bibliographic information published by the German National Library:

The German National Library lists this publication in the National Bibliography; detailed bibliographic data are available on the Internet at http://dnb.dnb.de .

Imprint:

Copyright © 2019 GRIN Verlag
Print and binding: Books on Demand GmbH, Norderstedt Germany
ISBN: 9783346148704

This book at GRIN:

https://www.grin.com/document/511686

Shoukat Ali R A

A Teacher's Guide on Complexometric Titration

GRIN Verlag

GRIN - Your knowledge has value

Since its foundation in 1998, GRIN has specialized in publishing academic texts by students, college teachers and other academics as e-book and printed book. The website www.grin.com is an ideal platform for presenting term papers, final papers, scientific essays, dissertations and specialist books.

A 'Student Handout-I' is a study cum laboratory material which is helpful to fresh teachers of the UG-PG departments as well as self studying students. The handout is also helpful for the students of both regular and distance education of post graduate departments. It eliminates the difficulties which are common in starting stage of teaching carrier related to solution preparations in various concentrations, calculations, procedures for the experiments totally the practical's set up. The material is also helpful to understand the role of reagents/chemicals used in the experiment, reaction conditions and structures. It is a preliminary effort we made to have ready material for the aspirants. The material will provide all the information related to the complexometric titrations with satisfactory. As we know no scientific corrections in the material, even if it has any corrections we welcome and correct in the next edition. Hope the 'Student handout-I' gain importance

Table of Contents

COMPLEXOMETRIC TITRATIONS

Definition:

Titrations in which an analyte (commonly divalent metal ions) is estimated by converting it into a suitable complex with the help of complexing agent (EDTA) in presence of other suitable reagents called complexometric titrations.

Basic principle:

The most versatile complexing agent is EDTA (ethylene diamine tetra acetic acid), even though it is a flexi dentate one, but it mainly acts as hexadentate ligand. It forms six co-ordinate complexes with most of the divalent metal ions [Ca Mg Cu Zn Ni Pb etc] via its two nitrogen atoms and four oxygen atoms.

During the formation of stable, 1:1 water soluble, colorless complex there will be the release of two H+ ions, as the number (n) of M-EDTA complex increases in the solution, there will be the removal of 2n[H+] ions, acidity increases in the solution. So we need to maintain the P^H of the solution with suitable basic buffer in order to have a M-EDTA complex. Hence complexometric titrations are generally carried out in basic P^H.

$$H_2Y^{2-} + M^{2+} = MY^{2-} + 2H^+$$

Indicator which show one color before complexation and give another colour after complexation with the metal called metallochromic indicator or complexometric indicator. In general indicator forms less stable complex with the metal (M) and EDTA can easily break the M-Indicator complex

Choice of indicators are based on

- Metal, its oxidation state and its stability with an indicator(Less stable)
- P^H of the solution
- Metal ion stability with EDTA (More Stable)

The stability order is

M-EDTA >/~ Indicator > M-Indicator

Hence,

Metal	Indicator choice	P^H Media
Mixture of $Ca^{2+} + Mg^{2+}$	EBT (ErioChrome Black-T)	Basic Buffer- basic P^H
Ca^{2+}alone estimation (in a Mixture of $Ca^{2+} + Mg^{2+})$	Patton and Reeders	KOH - basic P^H (also acts as masking agent to Mg^{2+})
Zn^{2+}	EBT	Basic Buffer- basic P^H
Cu^{2+}	Fast Sulphone Black -T	Conc. Ammonia- basic P^H
Pb^{2+}	Xylenol Orange	Hexamine- basic P^H
Ni^{2+} (Back estimation)	EBT (ErioChrome Black-T)	Basic Buffer- basic P^H

Chemicals Required:

1. 0.02M EDTA [Disodium Salt of EDTA-2° Standard]
2. 0.02M $ZnSO_4.7H_2O$ [1° Standard]
3. M^{+2} Stock Solution
   [Ca Mg Cu Zn Ni Pb etc]
4. Buffer
5. 8M KOH
6. Dil HNO_3

7. Hexamine

8. 1% Indicator/s etc...

Solution Preparations

a) For Solid Samples [Sl. No 1,2,3 and 5 in heading chemicals required]

Req. = Required

Weight req. = Req. Molarity × Molecular Weight × Req. Volume

1000

The formula can be applied to all solid chemicals, in molar concentrations.

Ex: If u wants to prepare 0.02M EDTA in 500 mL, then

Weight req. = $\dfrac{0.02 \times 372.24 \times 500}{1000}$

= .... g

b) Buffer solution : Dissolve 17.5g of ammonium chloride in 142mL of ammonia solution(Ammonium hydroxide) and dilute to 250mL

c) Indicator solutions: Generally 1% Indicator is used, dissolve 1g solid indicator in 100 mL of ethylene glycol or Alcohol.

Estimation of calcium and magnesium ions present in the given stock solution complexometrically by using EDTA solution

PROCEDURES

I Step: Standardization of EDTA solution:-

Pipette out 10mL of (0.02M) $ZnSO_4$ solution into a 250mL conical flask. Add 40mL of distilled water, 3mL of buffer and 3-4 drops of EBT indicator. Titrate the solution against EDTA solution (~0.02M) taken in the burette till color changes from wine red to blue. Repeat the titration to get concordant volumes.

II Step: Estimation of Calcium and Magnesium ions: -

Dilute the given stock solution using distilled water. Pipette out 10mL of stock solution containing calcium and magnesium ions into a 250mL conical flask. Add 40mL of distilled water, 3mL of buffer and 3-4 drops of EBT indicator. Titrate the solution against standardized EDTA solution taken in the burette till color changes from wine red to blue. Repeat the titration to get concordant volumes.

III Step: Estimation of Calcium ions alone: - Pipette out 10mL of given stock solution into a 250mL conical flask. Add 40mL of distilled water; add 5mL 8M KOH, and a pinch of Patten-Reeder's indicator. Titrate the solution against standardized EDTA solution until color change to blue from wine red. Repeat the titration to get concordant volumes.

Observations and calculations

a) Standardization of EDTA Solution

Burette: EDTA Solution
Conical flask: 10mL of $ZnSO_4$ + 40mL of distilled water + 4mL of buffer solution
        + 2-3 drops of indicator.
Indicator: Eriochrome Black - T
End Point: Wine red to Blue

Burette readings in mL	I	II	III
Initial reading in mL			
Final reading in mL			
Volume of EDTA Consumed in mL			

Concordant Value =.........mL

$(M \times V)_{EDTA} = (M \times V)_{ZnSO4}$

$(M \times V)_{EDTA} = \dfrac{(M \times V)_{ZnSO4}}{V_{EDTA}}$

b) Estimation of Calcium and Magnesium ions

Burette: Standardized EDTA Solution
Conical flask: 10mL of Stock Solution + 40mL of distilled water
+ 4mL of buffer solution + 2-3 drops of indicator
Indicator: Eriochrome Black T
End Point: Wine red to Blue

Burette readings in mL	I	II	III
Initial reading in mL			
Final reading in mL			
Volume of EDTA Consumed in mL			

Concordant Value = A.........mL

c) Estimation of Calcium alone:

Burette: Standardized EDTA Solution
Conical flask: 10mL of Stock Solution + 40mL of distilled water +5mL of 8M
KOH solution + 2-3 drops of indicator
Indicator: Patton and Reeders
End Point: Pink (red) to Blue

7

Burette readings in mL	I	II	III
Initial reading in mL			
Final reading in mL			
Volume of EDTA Consumed in mL			

Concordant Value = B.........mL

$(M \times V)_{Ca2+} = (M \times V)_{EDTA}$

$$M_{Ca2+} = \frac{(M \times V)_{EDTA}}{V_{ca2+}}$$

Estimation of Magnesium alone

Volume of EDTA Consumed by Mg^{2+} ions = (A-B)mL

$$M_{Mg2+} = \frac{(M \times V)_{EDTA}}{V_{Mg2+}}$$

For Ca^{+2} ions:-
Direct Method of calculation:-

Amount of Ca^{+2} ions present in given stock solution =

$$\frac{M_{stock\ solution} \times \text{Atomic weight of Ca [40.08]} \times \text{Reporting volume (100)}}{1000}$$

Conversion factor method of calculation:-

1000mL of 1M EDTA=40.08g of Ca^{+2}

1mL of 1M EDTA=0.04008g of Ca^{+2}

Amount of Ca^{+2} ions present in given stock solution =

$(MV)_{EDTA} \times \text{conversion factor [0.04008]} \times 10$

8

For Mg^{+2} ions-

Direct Method of calculation:-

Amount of Mg^{+2} present in given stock solution=

$$\frac{M_{\text{stock solution}} \times \text{Atomic weight of Mg [24.32]} \times \text{Reporting volume (100)}}{1000}$$

Conversion factor method of calculation:-

1000mL of 1M EDTA=24.32g of Mg^{+2} ions

1mL of 1M EDTA=0.02432g of Mg^{+2} ions

Amount of Mg^{+2} ions present in given stock solution=

$$(MV)_{\text{EDTA}} \times \text{conversion factor}[0.02432] \times 10$$

REPORT: -

Amount of calcium ions present in 100mL of stock solution ....

Amount of magnesium ions present in 100mL of stock solution ......

Estimation of copper ions present in the given stock solution complexometrically by using EDTA solution

PROCEDURES

I Step: Standardization of EDTA solution:-

Pipette out 10mL of (0.02M) $ZnSO_4$ solution into a 250mL conical flask. Add 40mL of distilled water, 3mL of buffer and 3-4 drops of EBT indicator. Titrate the solution against EDTA solution (~0.02M) taken in the burette till color changes from wine red to blue. Repeat the titration to get concordant volumes.

II Step: Estimation of Copper ions:- Dilute the given stock solution using distilled water. Pipette out 10 mL of stock solution containing copper into 250mL conical flask. Add 40mL of distilled water, 5mL of ammonia and a few drops of Fast Sulphone Black- F indicator. Titrate the solution against standardized EDTA solution until color change to blue from dark green. Repeat the titration to get concordant volumes.

Observations and calculations

   a) Standardization of EDTA Solution

Burette: EDTA Solution
Conical flask: 10mL of $ZnSO_4$ + 40mL of distilled water + 4mL of buffer
        solution + 2-3 drops of indicator.
Indicator: Eriochrome Black - T
End Point: Wine red to Blue

Burette readings in mL	I	II	III
Initial reading in mL			
Final reading in mL			
Volume of EDTA Consumed in mL			

                     Concordant Value =.........mL

$(M \times V)_{EDTA} = (M \times V)_{ZnSO4}$

$(M \times V)_{EDTA} = \dfrac{(M \times V)_{ZnSO4}}{V_{EDTA}}$

b) Estimation of Copper ions:

Burette: Standardized EDTA Solution

Conical flask: 10mL of Stock Solution + 40mL of distilled water
+ 5mL of Conc. ammonia Solution + indicator

Indicator: Fast Sulphone Black -T

End Point: Blue to dark green

Burette reading	I	II	III
Initial reading			
Final reading			
Volume of EDTA Consumed			

Concordant Value =.........mL

$(M \times V)_{Cu^{2+}} = (M \times V)_{EDTA}$

$M_{Cu^{2+}} = \dfrac{(M \times V)_{EDTA}}{V_{Cu^{2+}}}$

Direct Method:-

Amount of copper present in given stock solution =

$\dfrac{M_{stock\ solution} \times Atomic\ weight\ of\ Cu\ [63.54] \times Reporting\ volume\ (100)}{1000}$

Conversion factor method-

1000mL of 1M EDTA = 63.54g of Cu^{+2} ions

1mL of 1M EDTA =0.06354g of Cu^{+2} ions

Amount of Cu^{+2} ions present in given stock solution

$$= (MV)_{EDTA} \times \text{Conversion factor } [0.06354] \times 10$$

REPORT:- Amount of copper present in 100mL of the sample.......

Estimation of Lead ions present in the given stock solution complexometrically by using EDTA solution.

I Step: Standardization of EDTA solution:-

Pipette out 10mL of (0.02M) $ZnSO_4$ solution into a 250mL conical flask. Add 40mL of distilled water, 3mL of buffer and 3-4 drops of EBT indicator. Titrate the solution against EDTA solution (~0.02M) taken in the burette till color changes from wine red to blue. Repeat the titration to get concordant volumes.

II Step: Estimation of Lead ions: - Dilute the given stock solution using distilled water. Pipette out 10mL of stock solution of lead ions into 250mL conical flask, add 40mL of distilled water, 2-3 drops of xylenol orange indicator, add 3-4mL of dil nitric acid(till colour changes to yellow) and a pinch of powered hexamine. Titrate the solution against EDTA solution until color change from red to yellow. Repeat the titration to get concordant volumes.

Observations and calculations

a) Standardization of EDTA Solution

Burette: EDTA Solution

Conical flask: 10mL of $ZnSO_4$ + 40mL of distilled water + 4mL of buffer
solution + 2-3 drops of indicator.

Indicator: Eriochrome Black - T

End Point: Wine red to Blue

Burette readings in mL	I	II	III
Initial reading in mL			
Final reading in mL			
Volume of EDTA Consumed in mL			

Concordant Value =.........mL

$(M \times V)_{EDTA} = (M \times V)_{ZnSO4}$

$(M \times V)_{EDTA} = \dfrac{(M \times V)_{ZnSO4}}{V_{EDTA}}$

b) Estimation of Lead ions:

Burette: Standardized EDTA Solution

Conical flask: 10mL of Stock Solution + 40mL of distilled water indicator
+ 3-4mL of HNO_3(till yellow to red) + a pinch of hexamine

Indicator: Xylenol Orange

End Point: Red to Yellow

Burette reading	I	II	III
Initial reading			
Final reading			
Volume of EDTA Consumed			

Concordant Value =.........mL

$(M \times V)_{Pb^{2+}} = (M \times V)_{EDTA}$

$M_{Pb^{2+}} = \dfrac{(M \times V)_{EDTA}}{V_{Pb^{2+}}}$

Direct Method:-

Amount of Lead present in given stock solution =

$\dfrac{M_{stock\ solution} \times \text{Atomic weight of Pb } [207.2] \times \text{Reporting volume (100)}}{1000}$

Conversion factor method-

1000mL of 1M EDTA = 207.2g of Pb^{+2} ions

1mL of 1M EDTA = 0.2072g of Pb^{+2} ions

Amount of Pb^{+2} ions present in given stock solution

$= (MV)_{EDTA} \times \text{Conversion factor } [0.2072] \times 10$

REPORT:- Amount of Lead present in 100mL of the sample.......

Estimation of Nickel ions present in the given stock solution complexometrically by using EDTA solution.

I Step: Standardization of EDTA solution:-

Pipette out 10mL of (0.02M) $ZnSO_4$ solution into a 250mL conical flask. Add 40mL of distilled water, 3mL of buffer and 3-4 drops of EBT indicator. Titrate the solution against EDTA solution (~0.02M) taken in the burette till color changes from wine red to blue. Repeat the titration to get concordant volumes.

II Step: Estimation of Nickel ions: - Dilute the given stock solution using distilled water. Pipette out 10mL of stock solution of nickel ions into 250mL conical flask, add 25mL of standardized EDTA (allow to stand for 2 mins), add 40mL of distilled water, 4mL buffer and 2-3 drops of EBT indicator. Titrate the above solution against $ZnSO_4$ solution until color change from blue to wine red. Repeat the titration to get concordant volumes.

a) Standardization of EDTA Solution

Burette: EDTA Solution
Conical flask: 10mL of $ZnSO_4$ + 40mL of distilled water + 4mL of buffer solution
            + 2-3 drops of indicator.
Indicator: Eriochrome Black - T
End Point: Wine red to Blue

Burette readings in mL	I	II	III
Initial reading in mL			
Final reading in mL			
Volume of EDTA Consumed in mL			

Concordant Value =.........mL

$(M \times V)_{EDTA} = (M \times V)_{ZnSO4}$

$(M \times V)_{EDTA} = \dfrac{(M \times V)_{ZnSO4}}{V_{EDTA}}$

b) Estimation of Nickel ions:

Burette: $ZnSO_4$ Solution
Conical flask: 10mL of Stock Solution + 40mL of distilled water + 25mL of EDTA
            Solution + 4mL of Buffer Solution +2-3 drop of indicator.

Indicator: Eriochrome Black -T

End Point: Blue to Wine red

Burette reading	I	II	III
Initial reading			
Final reading			
Volume of $ZnSO_4$ Consumed			

Concordant Value =.........mL

Excess of EDTA (x) = $(V_{EDTA})/(V_{ZnSO_4})$ Standardization Step $\times$ $(V_{ZnSO_4.7H_2O})$Estimation Step

(x) =...........mL

V_{EDTA} Consumed for Ni ions = Volume of EDTA taken[25] – Excess of EDTA (x)

$(M \times V)_{Ni} = (M \times V)_{EDTA}$

$$M_{Ni} = \frac{(M \times V)_{EDTA}}{V_{Ni}}$$

Direct Method-

Amount of Nickel present in given stock solution=

$$\frac{M_{Ni} \times \text{Atomic weight of Ni [58.69]} \times \text{Reporting volume (100)}}{1000}$$

Conversion factor method-

1000mL of 1M EDTA =58.69 of Ni^{+2} ions

1mL of 1M EDTA =0.5869g of Ni^{+2} ions

Amount of Ni^{+2} ions present in given stock solution=

$(MV)_{EDTA} \times$ Conversion factor [0.05869] $\times 10$

REPORT:

Amount of nickel ions present in the given sample........

ROLE OF REAGENTS

EDTA:
Complexing agent, Hexa-dentate ligand,
It forms 1:1 water soluble, colorless complex

Basic Buffer:
It is a combination of ammonia and ammonium chloride
Used to maintain the basic P^H [~10]

KOH:
It maintains the basic P^H,
Used in the estimation Ca alone in presence of Mg
It acts as masking agent for Mg^{+2}
It selectively converts Mg as $Mg(OH)_2$
Because Mg has high polarizing power (charge/size) than Ca

Distilled water:

Used for dilution, upon dilution ions will move faster
In dilute solution we can observe proper color change

Inferences:

1. In case of Ca^{2+}, Mg^{2+}, Zn^{2+}, Cu^{2+} and Pb^{2+} ions we can follow direct titration methods. As the Metal(M)-(In)Indicator complex forms is less stable complex. EDTA can easily break these M-In complex and form M-EDTA complex

2. M-In complex is wine red in color and free ionized indicator is blue in colour.. where M= Zn , Ca&Mg , In= EBT

17

3. In case of Ni^{+2} , as it forms more stable complex with EBT, EDTA cant break the bond between M-In, so in the estimation of Ni we follow back titration.

Structures of disodium salt of EDTA and M-EDTA complex

Structure of Eriochrome Black – T

Structure of Patton-Reeder Indicator

Structure of Fast Sulphone Black – F

References:

- College Practical Chemistry – V K Ahulwalia
- Inorganic Practical Chemistry- Kaza Somashekara Rao
- Vogel's text book of quantitative chemical analysis- Jeffery et. Al
- Inorganic Chemistry- Puri Sharma Kalia